BEI GRIN MACHT SICH IHR WISSEN BEZAHLT

- Wir veröffentlichen Ihre Hausarbeit, Bachelor- und Masterarbeit

- Ihr eigenes eBook und Buch - weltweit in allen wichtigen Shops

- Verdienen Sie an jedem Verkauf

Jetzt bei www.GRIN.com hochladen und kostenlos publizieren

Schmitt, Theiss, Wache, Fuesers, Andree, Eichler, Eckler

Fluorescence spectrocopy of excitation energy transfer processes in Acaryochloris marina

GRIN Verlag

Bibliografische Information der Deutschen Nationalbibliothek:

Die Deutsche Bibliothek verzeichnet diese Publikation in der Deutschen National-
bibliografie; detaillierte bibliografische Daten sind im Internet über http://dnb.d-
nb.de/ abrufbar.

Impressum:

Copyright © 2007 GRIN Verlag GmbH
Druck und Bindung: Books on Demand GmbH, Norderstedt Germany
ISBN: 978-3-640-77014-4

Dieses Buch bei GRIN:

http://www.grin.com/de/e-book/159775/fluorescence-spectrocopy-of-excitation-
energy-transfer-processes-in-acaryochloris

www.zig-berlin.de

ZiGprint

2007-02

Investigation of metabolic changes in living cells of the Chl d- containing cyanobacterium Acaryochloris marina by time- and wavelength correlated single-photon counting

Franz-Josef Schmitt
Christoph Theiss
Karin Wache
Justus Fuesers
Stefan Andree
Hans Joachim Eichler
Hann-Jörg Eckert

ISSN 1862 4871

ZiGprint

2007-02

Investigation of metabolic changes in living cells of the Chl d-containing cyanobacterium Acaryochloris marina by time- and wavelength correlated single-photon counting

Franz-Josef Schmitt
Christoph Theiss
Karin Wache
Justus Fuesers
Stefan Andree
Hans Joachim Eichler
Hann-Jörg Eckert

Berlin, April 2007

Investigation of metabolic changes in living cells of the Chl *d*- containing cyanobacterium *Acaryochloris marina* by time- and wavelength correlated single-photon counting

Franz-Josef Schmitt[1], Christoph Theiss[1], Karin Wache[1], Justus Fuesers[1], Stefan Andree[1], Hans Joachim Eichler[1] and Hann-Jörg Eckert[2]

[1]Optisches Institut – P 1-1, Technische Universität Berlin, Strasse des 17. Juni 135, D-10623 Berlin, [2]Max-Volmer-Laboratorium, Technische Universität Berlin, Strasse des 17. Juni 135, D-10623 Berlin

Content

Abstract

Time- and wavelength-resolved fluorescence spectroscopy is an appropriate tool for quantitative and non-invasive investigations of living cells. Short measurement times with low excitation light intensities are necessary to observe variations of the fluorescence due to changes in the metabolism of the sensitive biological organisms.
With new techniques the fluorescence dynamics can be monitored simultaneously in a broad spectrum during a very short measurement time. That provides information about the spectral differences of the fluorescence dynamics which can vary in correlation with the metabolic changes.

The interaction of the photosynthetic subunits and especially the mechanisms regulating the energy transfer are presently interesting and open fields in photosynthesis research.

The phototrophic cyanobacterium *Acaryochloris marina* contains membrane extrinsic PBP antenna complexes and mainly Chl *d* containing membrane intrinsic core antenna complexes which absorb light and transfer excitation energy to the reaction center.

The results of our studies suggest a fast excitation energy transfer kinetics of 20-30 ps along the PBP antenna of *A.marina* followed by a transfer with a time constant of about 60 ps to Chl *d*.

Very often cells or cell fragments are kept at low temperature to decelerate ageing processes. Living cells of *A. marina* which were stored at 0°C for some time showed a reduced excitation energy transfer from the PBP to the Chl *d* antenna, which partially recovered when the sample had been kept at 25 °C for a short time.

The reduction of the excitation energy transfer might be caused by a mechanism that decouples the PBP antenna under cold stress conditions avoiding photo damage of the reaction center of PS II.

Keywords: Acaryochloris marina, fluorescence dynamics, spectroscopy, phycobiliprotein, chlorophyll d, fluorescence lifetime, excitation energy transfer, phycocyanin, allophycocyanin, cell metabolism, photosynthesis research, biotechnology, biomedicine, life sciences

Veränderungen im Metabolismus lebender Zellen des Chl *d-* haltigen Cyanobakteriums *Acaryochloris marina* untersucht mittels zeit- und wellenlängenkorrelierten Einzelphotonenzählens

Zusammenfassung

Zeit- und wellenlängenaufgelöste Fluoreszenzspektroskopie ist ein geeignetes Verfahren zur quantitativen und nichtinvasiven Untersuchung lebender Zellen. Kurze Messzeiten bei niedrigen Anregungsintensitäten sind notwendig, um Veränderungen der Fluoreszenz, verursacht durch metabolische Änderungen in empfindlichen biologischen Organismen, zu beobachten.

Mit moderner Technik kann die Fluoreszenzdynamik simultan in einem breiten Spektrum innerhalb einer kurzen Messzeit abgebildet werden. Dies liefert Informationen über spektrale Unterschiede der zeitabhängigen Fluoreszenz, welche in Korrelation zu metabolischen Veränderungen variieren kann.

Die Wechselwirkung der photosynthetischen Untereinheiten und insbesondere die Regulationsmechanismen des Energietransfers sind gegenwärtig interessante und offene Fragen der Photosyntheseforschung.

Das phototrophe Cyanobakterium *Acaryochloris marina* enthält membranextrinsische PBP Antennenkomplexe und überwiegend Chl *d* haltige membranintrinsische Core-Antennenkomplexe, welche Licht absorbieren und die Anregungsenergie zum Reaktionszentrum weiterleiten.

Die Ergebnisse unserer Untersuchungen weisen auf einen schnellen Energietransfer entlang der PBP Antenne mit einer Kinetik von 20-30 ps hin, auf welchen ein Transfer zum Chl *d* mit einer Zeitkonstanten von etwa 60 ps folgt.

Sehr oft werden Zellen oder Zellfragmente gekühlt, um Alterungsprozesse zu verlangsamen. Lebende Zellen von *A. marina*, die einige Zeit bei 0°C gelagert wurden, zeigten einen reduzierten Anregungsenergietransfer von der PBP Antenne zum Chl *d*, welcher sich innerhalb kurzer Zeit teilweise regenerierte, nachdem die Probe auf 25 °C erwärmt worden war.

Dieser Reduktion des Anregungsenergietransfers könnte ein Mechanismus zu Grunde liegen, welcher die PBP Antenne unter Kältestress entkoppelt, um eine lichtinduzierte Zerstörung des Reaktionszentrums im PS II zu verhindern.

1. Introduction

Acaryochloris marina which was discovered in 1996 has a unique composition of the light harvesting system. The chlorophyll (Chl) antenna of Photosystem II (PS II) contains mainly Chl *d* instead of the usually dominant Chl *a* and the Phycobiliprotein (PBP) antenna has a simpler rod shaped structure than in typical cynobacteria [1]. The energy transfer processes are still not fully explained and further spectroscopic studies are necessary to answer the open questions.

During the past 20 years there has been a remarkable growth in the use of fluorescence techniques in biological sciences. Quantification of toxic substances or health promoting components in fruits and vegetables, environmental monitoring, clinical chemistry and DNA sequencing are just a few areas of application [2].

With fluorescence spectrometry the wavelength- and also the time-dependency of the emitted light can be used to identify and quantify fluorescent substances.

In photosynthesis research information on excitation energy transfer among antenna pigments, charge separation within the reaction centers, forming of channels for non-photochemical quenching and the pigment–pigment or pigment-protein coupling can be gathered from analyses of the time decay and wavelength dependence of the emitted fluorescence [3],[4].

In living cells changes in the fluorescence emission can be detected that correspond to the cell metabolism. The influence of stress factors like cold, heat, high light intensities or the starvation of essential components alters the conformation of the coupled pigment-protein complexes. In order to take full advantage of the information content of the fluorescence emission, it is necessary to monitor the time- and the wavelength dependency of fluorescence light with sufficient resolution because the broad fluorescence emission bands are the sum of the fluorescence of hundreds of pigments in different protein environments. Therefore the information about the electronic structure of a single pigment bound to a protein cannot be resolved. Mathematical data analysis and the theory of optical spectra which today delivers tools to calculate the absorption of pigment-protein complexes are helpful to extract information of specific molecules in a pool of fluorescing pigments [5].

When observing fast metabolic changes the signal has to be collected simultaneously in the time and wavelength domain, because the fluorescence decay at different wavelengths can only be compared if the time resolved fluorescence is measured at different wavelengths at the same time. Therefore fast acquisition of time resolved data in a broad spectral range is required which can be accomplished by new spectrometer systems.

In this study we use time- and wavelength-resolved single photon counting to investigate the excited states dynamics in living cells of the prokaryotic cyanobacterium *Acaryochloris marina*.

A.marina is a very special prokaryotic cyanobacterium. It is still the only known oxygenic photosynthetic organism containing Chl *d* as the dominant antenna pigment. Additionally Phycobiliproteins (PBPs) and small amounts of Chl *a* are present [1],[6]. Chl *d* has a formyl group on ring 1 of the porphyrin headgroup, in place of the vinyl group in Chl *a*, shifting the Q_Y absorption maximum to 696 nm in methanol, ~30 nm more to the red as compared to Chl *a* (665 nm in methanol). *A.marina* is therefore able to exploit the near infrared light that penetrates to the shady environment where it lives [7]. In whole cells of *A.marina*, the main red absorption band is observed at 714-718 nm [1]. The room temperature steady state fluorescence of *A. marina* exhibits a broad Chl *d* band at 724 nm [8]. The red shift in absorbance of Chl *d* relative to Chl *a* is equal to an electronic energy gap of ~100 mV. Therefore it has been questioned how Chl *d* is able to split water in PS II [9],[10],[11].

Absorption- and fluorescence-spectroscopic studies with time and wavelength resolution have clarified the pathway of excitation energy transfer on the molecular level [11]. Also the mechanism of energy transfer and water splitting process has been analyzed in detail [12]. Chl *d* was shown to be the primary donor of the reaction center (RC) of PS I in *A. marina* [13]. The nature of the primary donor of PS II in *A. marina* is still in discussion. Recent studies suggest a start of the primary charge separation from the accessory Chl pigment which is Chl *d* (Chl d_1). This Chl *d* molecule is stabilized by the so-called special pair which consists of Chl *a*. Then the charge is quickly localized at Chl *a*. Therefore both Chlorophyll types, Chl *a* and Chl *d* are essential for the photochemistry in PS II of *A.marina* [11].

Fig. 1 shows a scheme of the cell and the PS II antenna system of *A.marina*. The prokaryotic cells of *A.marina* are containing staples of the thylakoid membrane where the photosynthesis takes place. The membrane extrinsic antenna is represented by a PBP rod which is associated to the PS II core antenna containing Chl *d* [14].

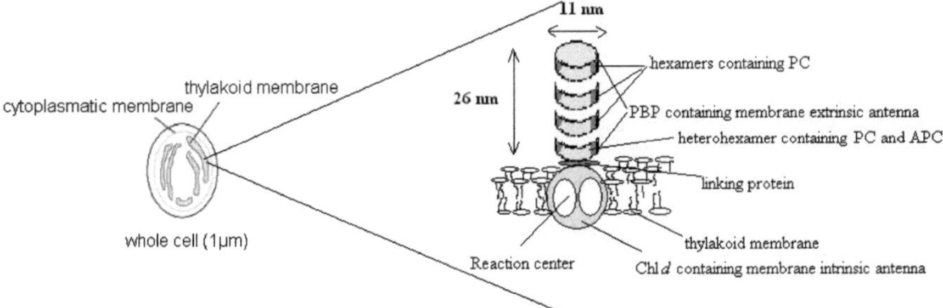

Fig. 1 Scheme of a cell of *A.marina* containing the thylakoid membrane (left side). At the right side the Light harvesting antenna complexes and Reaction center of PS II are shown according to Marquardt et al. [14]

The PBPs of *A.marina* have been reported to form aggregates of a simpler structure than those in typical cyanobacteria (fig. 1) [14]. They consist of four hexameric units, which resemble the peripheral rods of the typical cyanobacterial phycobilisomes (PBS) [14],[15]. Three of the hexamers were suggested to contain only phycocyanin (PC) and one to be a hetero-hexamer containing PC and allophycocyanin (APC). The excitation energy seems to be funneled directly from the APC-containing hetero-hexamer to Chl *d* of PS II without the involvement of an APC core as in typical cyanobacteria [15]. Isolated PBP aggregates of *A. marina* exhibit a fluorescence maximum at 665 nm (from APC) with a shoulder at about 655 nm (from PC) at room temperature [14]. This emission is also found in living cells of *A.marina* and the fluorescence decays with a time constant of 70 ps which is indicating the fast energy transfer from PBP to Chl *d* [16]. For reviews on fluorescence of *A. marina* and comparison with other systems, see Mimuro [17] and Itoh [18].

2. Materials and Methods

2.1 Algal culture

A. marina was grown as described in [19] in artificial sea water at 301 ± 2 K ($28° \pm 2$ °C) under an illumination intensity of 5 Wm^{-2} and continuous aeration. Before the measurements, the cells were spun down and gently re-suspended in a smaller amount of the growth medium to increase the cell density to about 5 µM of Chl *d*. This was done in order to obtain sufficiently high count rates during the fluorescence measurements while using low excitation intensities (1-5 Wm^{-2}) under continuous stirring necessary to avoid closing of the PS II reaction centers.

2.2 Excitation sources

For time resolved measurements a picosecond diode laser module was used for excitation at 632 nm (BHL-600, FWHM 60 ps, repetition frequency 20 MHz, Becker & Hickl GmbH, Berlin). The measurements were performed in a 3 x 10 mm cuvette shielded from room light. Fluorescence was detected at a right angle to the excitation beam. In order to suppress the scattered excitation light a long-pass emission filter was inserted between the cuvette and the detector (640ALP at an angle of 0° for 632 nm excitation (Omega Inc, cut-off wavelength 640 nm)).

In a modified setup to analyze the spectrum of the PBP emission, a LED was used for excitation at 600 nm (FWHM 800 ps, repetition frequency 8 MHz, Picoquant, Berlin). The long-pass emission filter (640ALP) was used at an angle of 30° (resulting in a cut-off wavelength 620 nm) giving additional information of the emission spectrum of PBPs in the range from 620 nm to 640 nm.

2.3 Fluorometer system

Fig. 2 shows two different types of single photon counting fluorometer systems with time- and wavelength resolution which are used for the fluorescence spectroscopic studies.

In Fig. 2 a) the system based on the technique of a delay-line detector is shown. During the measurement time- and space- correlation is used to determine simultaneously time- and wavelength information about the collected photons.

The double correlated single photon counting is achieved by using a microchannel plate photomultiplier (MCP-PMT) with delay-line anode (Europhoton GmbH, Berlin). In combination with a 120 mm crossed Czerny-Turner polychromator (MultiSpec, LOT) equipped with a 600 grooves/mm grating as a dispersive element, the space coordinate can be used for wavelength resolution [20].

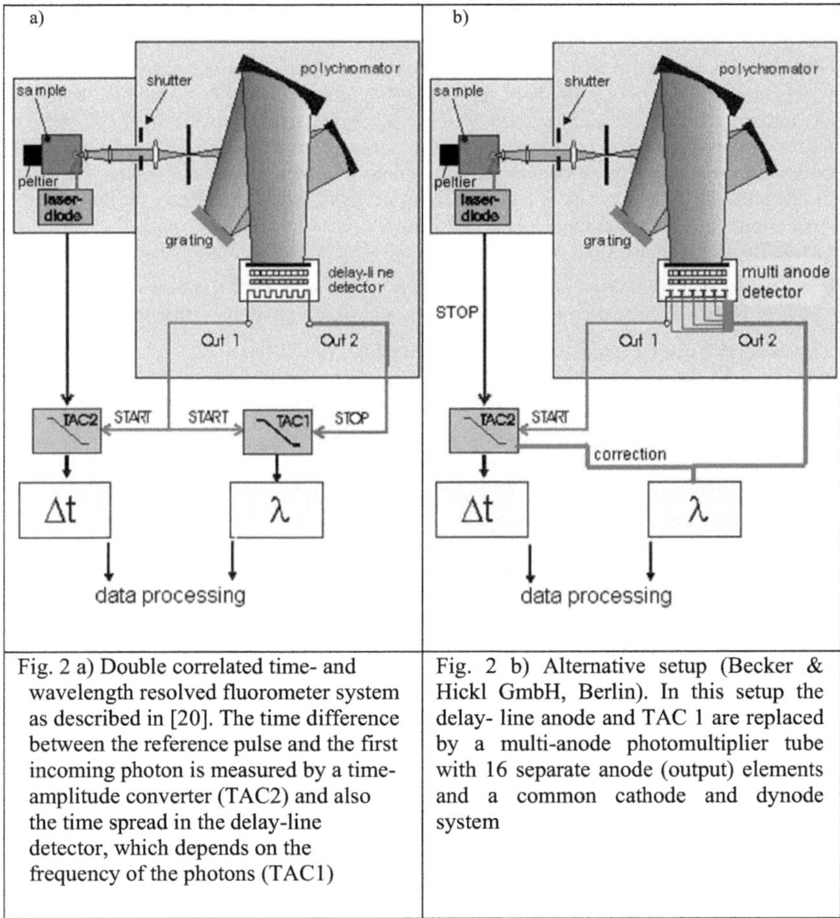

Fig. 2 a) Double correlated time- and wavelength resolved fluorometer system as described in [20]. The time difference between the reference pulse and the first incoming photon is measured by a time-amplitude converter (TAC2) and also the time spread in the delay-line detector, which depends on the frequency of the photons (TAC1)	Fig. 2 b) Alternative setup (Becker & Hickl GmbH, Berlin). In this setup the delay- line anode and TAC 1 are replaced by a multi-anode photomultiplier tube with 16 separate anode (output) elements and a common cathode and dynode system

As a function of the wavelength the fluorescence photons are deflected onto the photocathode of the MCP-PMT in the focal plane of the polychromator. A photoelectron is emitted at the inner side of the cathode and amplified by two microchannel plates producing a spatially limited electron cloud, which finally impinges on the delay-line anode. The electric charge moves to the opposite ends of the delay-line and arrives at both ends at different times depending on the position where the electron cloud hit the anode. The measured time difference between these two output pulses is then used to calculate the space-coordinate of the photon which allows the determination of its wavelength. Therefore this detector setup allows simultaneous monitoring of the time and wavelength dependence of fluorescence light with picosecond time resolution.

The two outputs of the delay-line anode are amplified by two 1GHz preamplifiers (Ortec 9306) and further processed by constant fraction discriminators (CFDs) (Tennelec TC 454). The Output of one of the CFDs provides the start signal for two time-to-amplitude converters

(TACs, Ortec 457 for space domain and Tennelec TC 864 for time domain). The stop signal for the space-domain TAC is provided by the output of the second CFD. A small fraction of the excitation laser light is reflected onto a fast photodiode that provides the stop signal for the time-domain TAC. The outputs of the two TACs are processed by a personal computer. The resulting data is stored in a two-dimensional channel matrix of size 256 x 1024, with 256 channels in the space domain corresponding to spectral resolution (wavelength coordinate) and 1024 channels in the time domain corresponding to temporal resolution (time coordinate of the fluorescence decay). The wavelength coordinate is provided directly by the output of the space-domain TAC. A correction for different photon wavelengths is applied to the output values of the time domain TAC in order to obtain the correct value of the time coordinate.

The technique of single photon counting operates with high signal to noise ratio which was $S/N \sim \sqrt{S} > 100$ in all measurements performed in this study. With the setup presented in Fig. 2a) this S/N ratio can be achieved in measurement times of < 10 min. for up to 30 wavelength sections of 5-10 nm spectral widths. The instrumental response function (IRF) of this system has 150 ps full width at half maximum (FWHM), limiting the time resolution to about 30 ps.

Fig. 2b) shows a time- and wavelength resolved multi anode detector system with 16 output (anode) elements (PML-16 , Becker&Hickl, Germany). Compared to the sensitive delay-line anode system the multi anode system yields a dramatically increased detection rate up to 1.000.000 counts/sec with the mentioned 20 MHz diode laser. The core of the PML-16 is a Hamamatsu R5900 16 channel multi-anode photomultiplier tube with 16 separate output (anode) elements and a common cathode and dynode system as described in [22]. The wavelength resolution of the multi anode system is limited by the number of anodes in comparison to the whole detected spectral bandwidth that is determined by the grating. Using the polychromator with a 600 grooves/mm grating the spectral bandwidth of the PML-16 is about 6 nm /channel.

New techniques using several photomultiplier tubes allow even higher count rates because pile up can be neglected if detection systems with detector arrays are used [23].
Measurements with higher time resolution were achieved employing a monochromator system (McPherson Instrument) with a MCP-PM-tube (Hamamatsu). The IRF of this system is 90 ps FWHM allowing a time resolution which is shorter than 20 ps. These measurements were performed in a time window with 4096 channels and 5 ns in total. Typical calibration values were 1.22 ps per channel in the time domain.

A Peltier cooling/heating system (Peltron GmbH, Germany) allows the choice of any temperature in the range of 260 K to 330 K in every setup. The regulation of the sample temperature is necessary to avoid warming up of the sensitive cells during the measurement. Furthermore the choice of different temperatures was necessary to investigate the influence of cold stress. For further details see also [21].

2.4 Data analysis

For a correct data analysis the knowledge of the response of the system to the laser pulse without fluorescence is necessary. The temporal width of this signal is caused by the laser pulse duration and additional electronic broadening processes. The deconvolution of the instrumental response function (IRF) with the fluorescence signal helps to better the time resolution down to 20 % of the laser pulse FWHM.

After every experiment the IRF was measured by detecting the attenuated excitation light scattered from a cuvette filled with distilled water using the same conditions as in measuring fluorescent samples. The fluorescence decay was analyzed employing Levenberg-Marquardt algorithm for the minimization of the reduced (χ_r^2) [24]. The algorithm was implemented using Matlab© software (The MathWorks Inc.). The decay was fitted to a multi-exponential decay model

$$F(t,\lambda) = \sum_{j=1}^{n} a_j(\lambda)\, e^{-t/\tau_j}, \tag{1}$$

with up to four components (n=4). The quality of the fit was judged by the value of χ_r^2 and by the degree of randomness of residuals (difference between the experimental data points and the fit). As an alternative to the Matlab implementation the software of Globals Unlimited (University of Illinois, Urbana, USA) providing equivalent analysis options was used (also see Gilmore [25]).

Fig. 3 a) shows a fluorescence measurement of whole cells of *A.marina* collected after excitation with 632 nm at room temperature. The number of the registered photons at each wavelength and each time channel was stored in a 2-dim. data matrix. In this matrix the lines contain 1024 entries which represent the 1024 time channels. The columns contain 256 entries representing the wavelength channels. In Fig. 3a) the data matrix is shown as a color intensity plot (CIP). A CIP is a plot of the fluorescence intensity (pictured by the color or greyscale) in dependency of wavelength (y-axis) and time (x-axis). The fluorescence decay curves of any wavelength section can be plotted from this matrix (Fig. 3 b).

Fig. 3 c) shows the fluorescence spectrum at different times. Plots of the spectra at different time intervals are called time resolved spectra. Time resolved spectra help to identify the fluorescence of certain fluorophores at later times when the fluorescence emission of other pigments or scattered light already decayed.

In fig. 3 c) the main emission is observed at 645 – 660 nm (PBP emission) immediately after excitation (0 ps) while one will find strongest fluorescence at 725 nm (Chl *d* emission) after 1 ns. For better visibility the spectrum after 1 ns is multiplied with a factor 5.

The fluorescence decay curves of different wavelength sections are fitted according to the multiexponential model (eq. 1).

The multiexponential fits of several decay curves are often fitted together (global fit) with common values of lifetimes τ_j (linked parameters) and wavelength-dependent pre-exponential factors $a_j(\lambda)$ (non-linked parameters). The result of such analysis is usually plotted as graph of $a_j(\lambda)$ for all wavelength independent lifetimes τ_j representing so-called decay associated spectra (DAS) showing the energetic position of individual decay components (fig. 3d).

For further details on the data analysis see also [16].

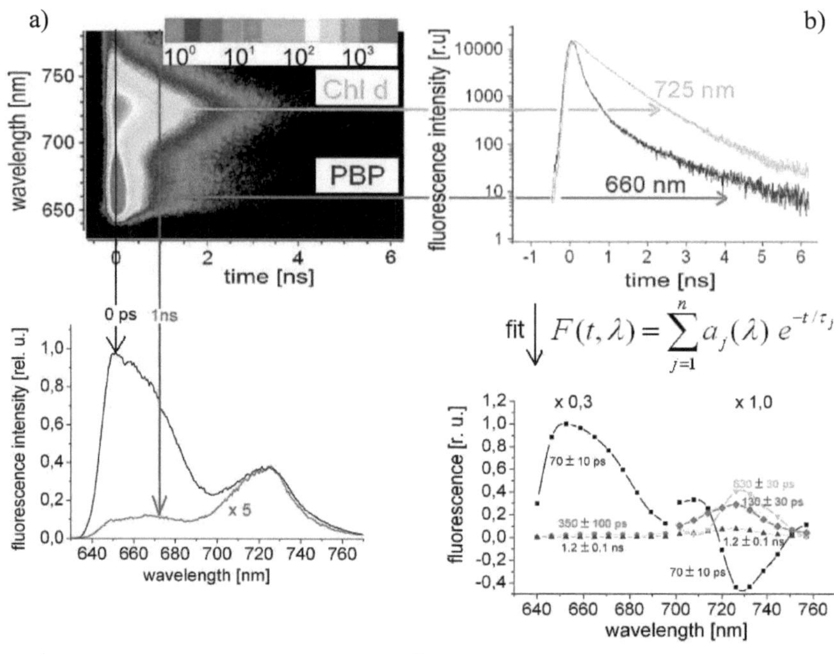

$$\text{fit} \Bigg\downarrow \quad F(t,\lambda) = \sum_{j=1}^{n} a_j(\lambda)\, e^{-t/\tau_j}$$

Fig. 3 a) Color Intensity Plot (CIP) of a measurement on *A.marina* after excitation at 632 nm at 298 K b) Fluorescence decay curves at 660 nm and 725 nm c) Fluorescence spectra at 0 ps and 1 ns (at 1ns multiplied with a factor 5) d) Decay associated spectra (DAS) of a global fit in the range 640 nm – 690 nm (multiplied with a factor 0,3) and a global fit in the range 700 nm – 760 nm.

3. Results

To analyze the results of a spectroscopic study on living cells of *A.marina* one should start with the interpretation of the time integrated fluorescence spectrum.

Fig. 3a) (top left) shows the color intensity plot (CIP) of the fluorescence emission of *A.marina* after excitation with 632 nm (pulsed laser, FWHM 60 ps) at room temperature. The CIP displays a broad emission of the PBP fluorescence (640-670 nm), which is not fully resolved because the long pass emission filter (640ALP) which is used to eliminate scattered laser light influenced also the fluorescence emission at shorter wavelengths than 640 nm. Therefore the fluorescence spectrum after excitation at 600 nm wavelength with a pulsed LED was used to analyze the pigment composition of the PBP antenna. After excitation at 600 nm the fluorescence can be resolved in the whole spectrum from about 610 nm – 750 nm.

Fig. 4 shows the time integrated fluorescence spectrum of *A. marina* after excitation at 600 nm. The spectrum is characterized by a broad emission peaking at 650 nm with a shoulder at

660 nm. An appropriate spectral fit with Gaussian emission bands requires at least three Gaussian bands located at 644±3 nm, 665±3 nm and 725±3 nm. Minor contributions are found around 703 nm and 740 nm.

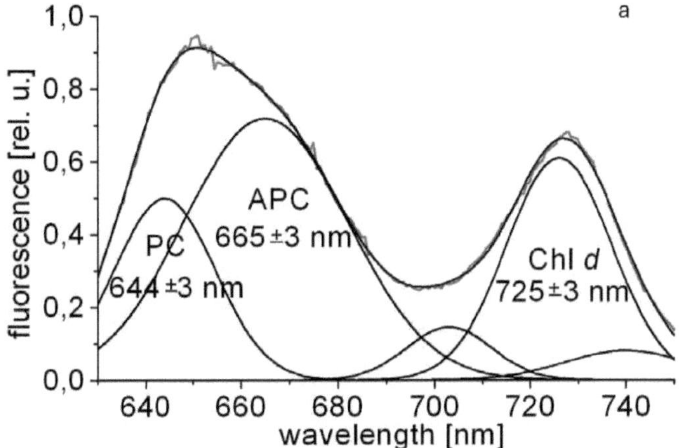

Fig. 4 Time integrated fluorescence spectrum with spectral Gaussian fits after excitation at 600 nm at room temperature

For analysis of the fluorescence decay the 632 nm diode laser was used because the pulse duration is short enough to resolve fluorescence decay components with 30 ps resolution. This is not the case with the 600 nm LED (FWHM 800 ps). After excitation with the 632 nm laser diode the 644 nm and the 665 nm emission bands appear as a broad emission with a maximum at 650 nm and a shoulder at 665 nm, both decaying predominantly with a lifetime of 70±10 ps at 298 K [16].

The DAS which represent the fit results of the wavelength dependent fluorescence decay curves deliver information of the coupling of the pigments and proteins additionally to the pigment composition.

Fig. 5 displays the results of a fit according to a multi-exponential model as described in section 2 (subsection data analysis). The decay curves of the wavelength sections in the PBP emission (640 – 690 nm) were fitted with linked lifetimes τ_j (global fit) and also the decay curves in the Chl d emission (700 – 760 nm) were fitted together with linked lifetimes.

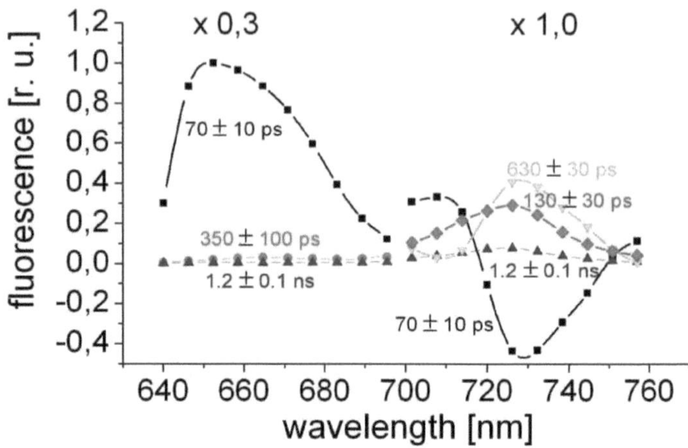

Fig. 5 DAS of the time resolved spectra after excitation with the 632 nm laser system at room temperature. The PBP fluorescence (640-680 nm) is scaled with a factor of 0,3. The small numbers at the plots denote the corresponding decay time.

Fig. 5 shows that the 725 nm fluorescence can be fitted with a four-exponential decay kinetics. Other studies showed that the lifetimes depend on the open or the closed state (F_0 or F_m) of the reaction center of PS II [16]. A lifetime with a time constant of 70±10 ps and negative amplitude was necessary to achieve an accurate fit quality of the Chl *d* fluorescence (724 nm). Such a negative amplitude of an exponential decay describes a fluorescence rise in the time.

If only three decay components are used for a local fit (only one wavelength section) of the Chl *d* fluorescence (725 nm) the value of χ_r^2 which is describing the fit quality (see section 2 subsection data analysis) is about 1.2 in average. With only three decay components the 70 ps component in the Chl *d* fluorescence (725 nm) can not be resolved.

Fig. 6 shows the distribution of the residuals after fitting the fluorescence decay in only one wavelength section at 725 nm (Chl *d* fluorescence). The residuals of a three component fit are shown at the left side. If only three decay components are used the residuals show systematic aberrations in the fit of the first 100 ps (red circle).

The right panel shows that this systematic aberration vanishes if a fourth decay component is added. Then the fit results show the existence of a component with a time constant of 70 ps which is near to the resolution limit and with negative amplitude in the Chl *d* area (725 nm). Such a component is called "fluorescence rise term" and is typical if an energy transfer to the investigated fluorescence emitter (here: Chl *d*) occurs.

Therefore very detailed analysis of the fit residua is necessary to resolve the 70 ps component in the Chl *d* fluorescence (725 nm) because the absolute value of χ_r^2 which is widely used for estimating the fit quality does not change much if a fourth decay component is added in comparison to a three component fit but the residua are more randomized if four decay components are used. Therefore the degree of randomness of the residua should be used for estimating the fit quality and not the absolute value of χ_r^2 .

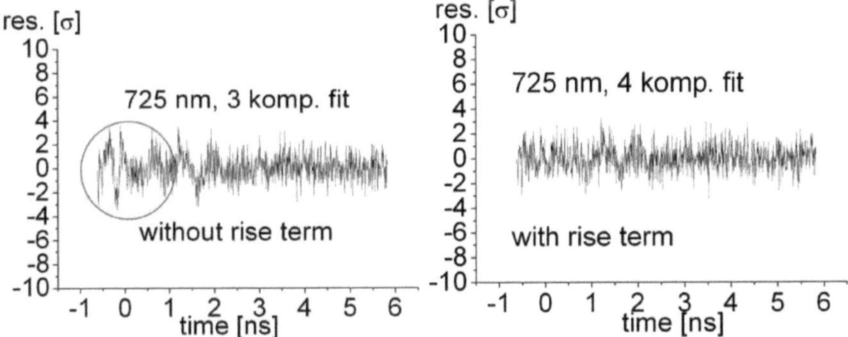

Fig. 6 Distribution of the residua after fitting the fluorescence decay in only one wavelength section at 725 nm (Chl *d* fluorescence).

It was not possible to achieve a similar good fit quality in the BPB emission band (645 nm to 665 nm). The value of χ_r^2 is about 1.6 and no better results were obtained by adding further decay components. The aberrations of the residua were similar to those of the 3 component fit of the Chl *d* fluorescence (725 nm) as shown in fig. 6 (left panel).

The reason for these systematic aberrations in the PBP fluorescence (645- 665 nm) can be understood. The structure of the light harvesting system of *A.marina* suggests that fast energy transfer steps occur between different PC molecules in one hexamer and also between the different hexameric structures of the PBP rod or between PC and APC in the heterohexamer (fig. 1). All these energy transfer steps should be very fast in comparison to the resolution limit (<30 ps) because the distance of the chromophores is very short in the PBP rods [26]. According to simulations at the Optical Institute of the TU Berlin by Justus Fuesers (data not shown) several fluorescence decay and fluorescence rise components with fast time constants which can not be resolved are expected in the PC and APC (640 nm to 670 nm) fluorescence of the PBP rod of *A.marina*. Therefore an accurate fit without any aberrations requires a lifetime distribution with very fast decay components in the PBP fluorescence (645 – 665 nm). This was not done in this experiment.

Fig. 7 displays a better resolved DAS of the PBP fluorescence after excitation of whole cells of *A.marina* with 632 nm at room temperature. The fluorescence measurement was done with the described monochromator system (McPherson Instruments) with a better time resolution of 20 ps. (see section 2, subsection fluorometer system).

The results of a three component fit showed a decay component with 30±10 ps with positive amplitude in the range from 645 to 655 nm and negative amplitude in the range from 655 to 665 nm. Other experiments clarified that the 30 ps rise term (negative amplitude) is also found at 680 nm (data not shown).

In this high resolved experiment shown in fig. 7 decay components with time constants > 1ns could not be resolved because the whole window length of the time measurement was just 5 ns. It was sufficient to fit the PBP fluorescence with three decay components as shown in fig. 7. The value of χ_r^2 was 1.2. Therefore the additional resolution and fit of a component of 30 ps resulted in an improvement of the χ_r^2 from 1.6 (without 30 ps component) to 1.2 (with 30 ps component).

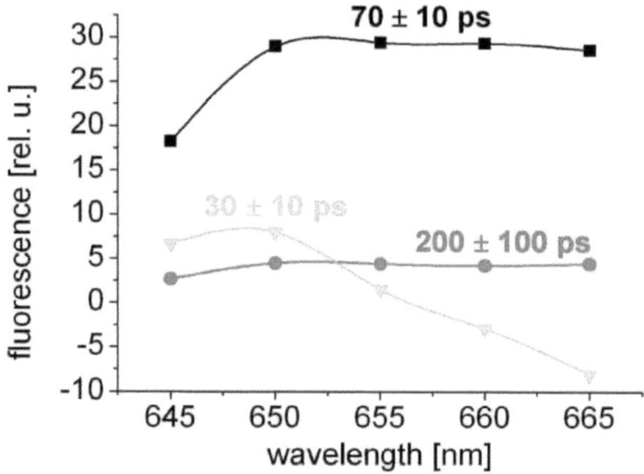

Fig. 7   DAS of the PBP fluorescence of *A.marina* after excitation with 632 nm at room temperature. With a time resolution of 20 ps a time constant of 30 ps could be resolved in addition to the results shown in fig. 5

A similar time constant of about 20 ps was also found in preliminary ps-pump-probe absorption measurements on isolated PBP-aggregates of *A.marina* by C.Theiss, C. Cardenas Chavez and S.Andree (unpublished results).

4. Decoupling of the PBPs under cold stress

The regulation of cell metabolism is a topic of high interest in biology, medicine and pharmacy. Some metabolic changes occur on very short time scales and therefore it is difficult to observe the changes. Many microscopic techniques, which are used for investigations need vacuum or cryostatic temperatures. Therefore these techniques are not adequate to investigate metabolic changes in vivo. Fluorescence spectroscopy is very sensitive and can therefore work with low light intensities measuring the sample in vivo in solution at room temperature.

Excitation energy transfer from PBP to PS II has been reported to be regulated in cyanobacteria [27] and changes of the transfer efficiency have been observed under stress conditions like cold stress [28]. The mechanisms regulating the energy transfer are still unknown and therefore a topic of the recent research. It was suggested that the regulation of the energy transfer helps to protect the PS II from light induced damage. Therefore we investigated the effect of cold stress on the fluorescence dynamics in *A.marina*.

Figure 8 shows the change of the fluorescence emission of whole cells of *A.marina* after incubation at 273 K (0°C). The cells were excited with the 632 nm diode laser.

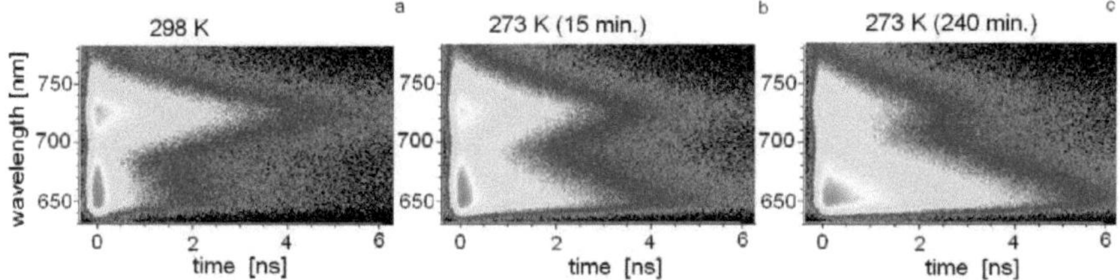

Fig. 8   CIPs of the fluorescence of whole cells of *A.marina* stored at room temperature (left), stored at 273 K for 15 minutes (middle) and 4 hours (right), after excitation with 632 nm.

The CIPs in fig. 8 show that the fluorescence in the PBP emission (645 – 665 nm) was decaying more slowly after incubation of the sample at 273 K. Simultaneously the intensity of the Chl *d* fluorescence was decreasing. The latter effect is clearly visible in the time integrated fluorescence spectrum as shown in fig. 9.

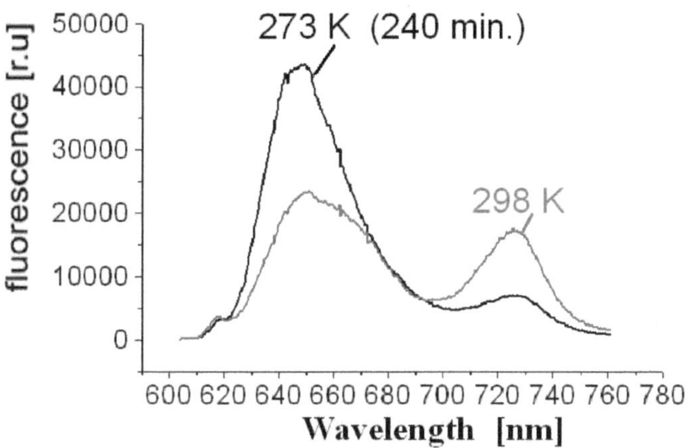

Fig. 9   Time integrated fluorescence spectrum at 298 K and after 240 min. of incubation at 273 K

Fig. 9 shows that the time integrated fluorescence of the PBP emission (645 – 665 nm) was nearly equal to the Chl *d* emission (725 nm) at 298 K. The time integrated fluorescence of Chl *d* (725 nm) was decreasing dramatically in comparison to the PBP fluorescence (645- 665 nm) after the incubation of living cells at 273 K for 4 h. The reason for the rise of the time integrated fluorescence in the PBP area (645 – 665 nm) was the deceleration of the fluorescence decay in time.

Fig. 10 displays the changes in the fluorescence dynamics at 660 nm.

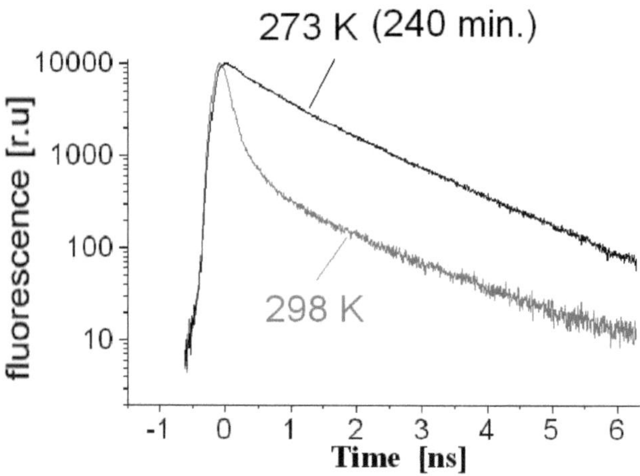

Fig. 10 Fluorescence decay at 660 nm at 298 K and after 240 min. of incubation at 273 K. After the incubation at 273 K the PBP fluorescence decay was much slower than at room temperature

The decay of the fluorescence at 660 nm (Fig. 10) showed the strong deceleration of the PBP fluorescence which appeared simultaneously to the decrease of the Chl *d* fluorescence. It is visible in fig. 8 that the effects of the PBP deceleration and the simultaneous decrease of the Chl *d* fluorescence were not appearing spontaneously but both effects increased during a time interval of up to four hours.

Fig. 11 and Fig. 12 show the quantitative analysis of the time dependency of the PBP fluorescence (Fig. 11) and the simultaneous decrease of the Chl *d* fluorescence (fig. 12). Fig. 11 displays the relative contribution of the decay components in the APC fluorescence emission (665 nm). It shows that the contribution of the 70 ps fluorescence component decreased from 98 % (298 K) to 25 % (4 h at 273 K). Complementary to this decrease the amplitude of a 600 ps component and a 1.4 ns component were increasing. APC is assumed to emit the long wavelength range (maximum at 665 nm) of the PBP fluorescence. The time constant of the 70 ps component decelerated to 140 ps (data not shown) when the sample was cooled for 4 h. In fig. 11 an average value of 100 ps is shown and a decrease of the amplitude of this 100 ps component is clearly visible. The time constant of the medium component which is denoted with 600 ps was varying very strong between 200 ps and 900 ps (data not shown). Fig. 11 displays an average value of 600 ps.

After rewarming the cooled sample from 273 K to 298 K the amplitude of the 1.4 ns component reduced again and the 100 ps component recovered partially. Only the 600 ps component remained unaffected and did not significantly change after warming up the cells.

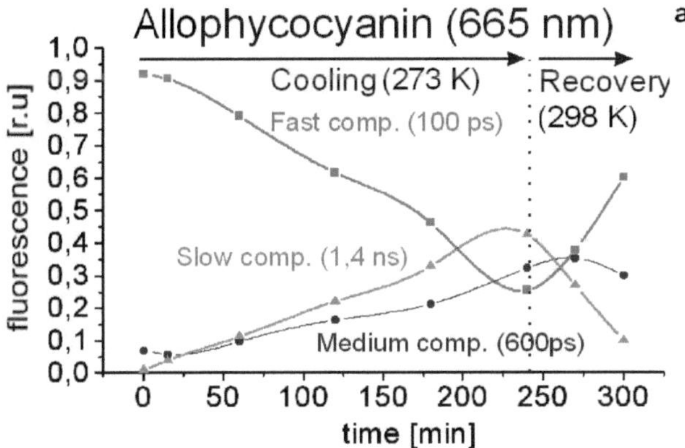

Fig. 11 Initial amplitudes of the PBP fluorescence decay components at 665 nm after excitation with 632 nm in dependency of the incubation time of *A.marina* at 273 K.

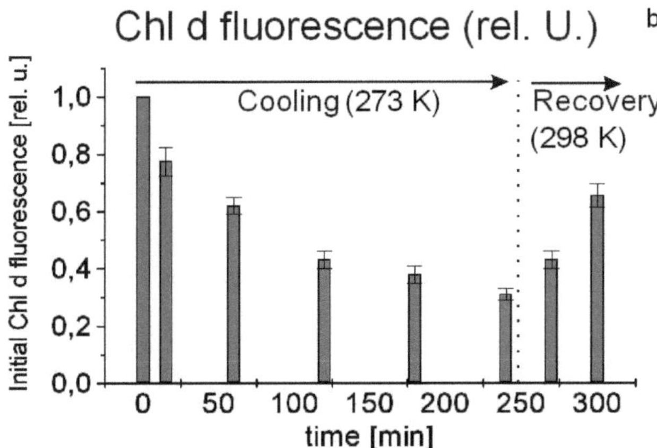

Fig. 12 Initial amplitude of the whole Chl *d* fluorescence (all decay components) which corresponds well with the amplitude of the fast fluorescence decay component (100 ps) in the PBP emission (fig. 11).

Fig. 12 shows the initial amplitude of the Chl *d* fluorescence during the cooling and subsequent rewarming. The time constants of the Chl *d* emission did not vary in a significant way (data not shown). The plot of the Chl *d* emission in fig. 12 fits the shape of the red curve of the 100 ps component shown in fig. 11 with small aberrations.

It is an interesting question whether the decoupling of the PBP antenna system is a general protection mechanism. In that case also other stress conditions should lead to similar decoupling effects. Therefore we measured the effect of heat stress and we observed a similar decoupling of the PBP antenna at 50°C (data not shown). Stress from strong light seemed to amplify the heat induced decoupling of the PBPs in *A.marina*.

There exist similar decoupling effects observed at different stress conditions in different PBP containing species. For example cells of Synechocystis sp. which have been starved for nitrogen compounds start to degenerate the PBP antennae (phycobilisomes) during 24 h while illuminated. A gene termed nblA plays a key role in PBP degradation [29]. NblA mutants did not show the PBP degradation but they also did not show any growth disadvantage anyway. Our measurements showed that these nblA mutants who are not able to decompose the PBP antenna systems are able to decouple their whole phycobilisomes. This is visible in the PBP fluorescence decay of the Synechocystis mutants after 24 h (data not shown) and similar to the deceleration of the PBP fluorescence appearing in *A.marina* at cold stress conditions after 1 h as shown in fig. 11.

5. Excited state populations according to a model of rate equations

To calculate the transfer dynamics between emitting and non-emitting species in the antenna systems one can assume a simple model of linear differential equations with constant coefficients. Such rate equation systems help to describe the dynamics of excited state populations of different coupled fluorophores. Therefore the assumptions and equations of this model are briefly presented.

The fluorescence of a single molecule without energy transfer processes is decaying mono-exponentially in time assuming the most simple model with a time constant depending on the sum of positive rate constants, each describing the rate of any relaxation process (fluorescence, inter system crossing or energy dissipation) [2] ,[30].

Energy transfer between two coupled pigments might occur from a donor molecule to an acceptor but also backwards from the acceptor to the donor. If energy transfer is possible, then the fluorescence decay has to be calculated from the solution of a set of coupled linear differential equations 1st order (one equation for each pigment in an energy transfer chain of an arbitrary number of n coupled molecules) [21],[30].

This set is solved by a sum of exponential functions with different time constants. In a non-degenerated set with n pair wise different eigenvalues of the transfer (imaging) matrix T describing the change $\dot{N}$ of all n state population densities N the set of differential equations can be written like:

$$\dot{N} = TN \tag{2}$$

The solution for the ith state population density is given by

$$N_i(t) = \sum_{j=1}^{n} U_{ij} e^{\gamma_j t} \tag{3}$$

where γ_j is denoting the jth eigenvalue and U_{ij} the ith component of the jth eigenvector of T .

Equation (3) is proportional to the assumed fluorescence decay of equation (1) with $\gamma_j = -(\tau_j)^{-1}$ according to the proportionality between fluorescence intensity and excited state population density.

Transforming the eigenvector matrix U and the eigenvalues in the diagonal matrix Γ to the transfer matrix T

$$T = U\Gamma U^{-1} \tag{4}$$

one can calculate the transfer matrix from the observed fluorescence intensities and fluorescence decay time constants.

Usually one does not find all eigenvector components (i.e. when some state populations do not show fluorescence or when not all fluorescence decay components can be resolved). So equation (4) can not be used to find a unique transfer matrix T. But a reasonable assumption of T with iterative comparison of the eigensystem (U,Γ) with the fluorescence decay helps to find a solution for T which is consistent with the globally observed fluorescence dynamics in all wavelength sections according to equation (1) [30]. Such an iterative comparison of the measurement data (eigenvectors and eigenvalues) and a reasonable transfer matrix T can be used to calculate a suggestion for the transfer rates of a system of arbitrary complexity which leads to the observable fluorescence decay. The iterative comparison can be achieved by using various simple algorithms.

For an example one could start with an arbitrary reasonable transfer matrix T_{Start}. Then the eigenvectors and eigenvalues of T_{Start} are calculated. In the first step one will set the observed decay times for all eigenvalues $\gamma_j = -(\tau_j)^{-1}$. Therefore the eigenvalues are changed but the eigenvectors are not. Doing the transformation according to (4) with the changed (experimentally observed) eigenvalues one will find an improved suggestion T for the transfermatrix T_{Start}.

There exist some constraints for the transfer matrix T. For an example the rate constant for the decay rate of pigment one according to an energy transfer from the pigment one to the pigment two has to be the same as the fluorescence rise rate of pigment two according to this energy transfer. Therefore all violated constraints have to be corrected in the transfer matrix T. After that one can calculate the eigenvectors and eigenvalues of T again. One can set the observed decay times for all eigenvalues $\gamma_j = -(\tau_j)^{-1}$ again. Doing the transformation according to (4) with the changed (experimentally observed) eigenvalues one will find a further improvement of T.

The described algorithm converged for calculating the transfer rates in *A.marina*. There exist several other connections between experiment and calculation, which could be implemented in the algorithm. For example one could also use the observed amplitudes of the different fluorescence components to correct the eigenvectors of T_{Start} instead of the eigenvalues.

Using this algorithm with 5 iterations, for *A.marina* the couplings of fig. 13 were found, describing the room temperature measurements of whole cells (298 K).

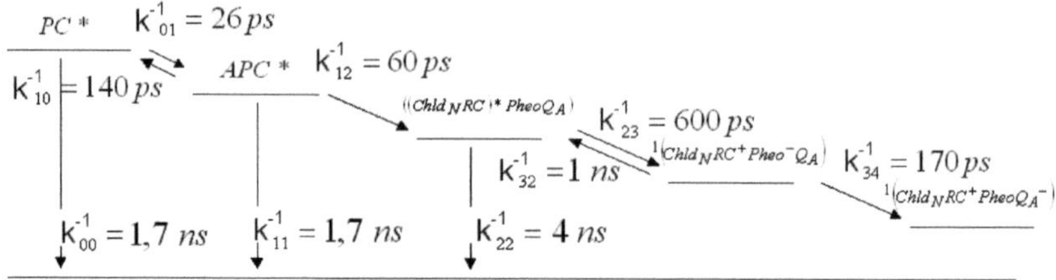

Fig. 13 Assumption of rate constants between different states of the primary photosynthetic processes in the Antenna systems and reaction center of PS II of *A.marina* , which would result in the fluorescence DAS shown in fig.4b).

Fig. 13 is not necessarily the only solution for the observed fluorescence pattern, but it can be shown that the rate constants k_{01}, k_{10} and k_{12} describing the dynamics of the PBP fluorescence are necessary to explain the observed 20 ps dynamics according to fig. 7 and the 70 ps component shown in fig. 5 and 7. We show a solution for a 20 ps fluorescence term in the PBP fluorescence because the results of C.Theiss, C. Cardenas Chavez and S.Andree (unpublished results) showed a 20 ps excited state density dynamics instead of the 30 ps component shown in fig. 7 and the time resolution of C.Theiss et al. was much better (400 fs).

Calculating the expected fluorescence decay from this coupled system (fig. 13) one will find the DAS shown in fig. 14, which is in agreement with the measurements shown in fig. 5 and 7. The spectral shape of the emission bands in fig. 14 is determined from the experimental data with Gaussian fits comparable to the fit results in fig. 5 (data not shown). After calculating the DAS from fig. 13 a correction of the 130 ps fluorescence at 725 nm was done, because one observes also PS I fluorescence at 725 nm which is not incorporated in the calculations for the PS II presented here [16].

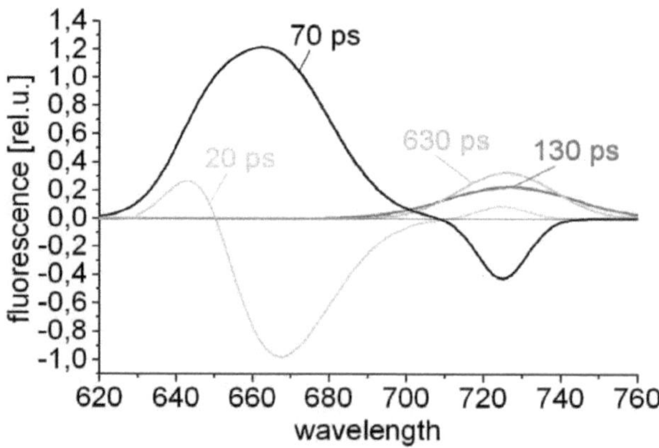

Fig. 14   DAS calculated from the transfer rates shown in fig. 9 after adding the PS I fluorescence.

The expected fluorescence decay of the coupled system presented in fig. 13 is in good agreement with the experimental data. Comparing the calculated DAS (fig. 14) with the measurement fit presented in fig. 5 the very similar dominating 70 ps decay in the PBP fluorescence (640 – 680 nm) is found. The 20 ps component is describing the energy transfer from PC to APC as presented in fig. 7. Also the composition of the Chl d fluorescence (725 nm) is calculated in very good agreement to the measurement (fig. 5).

The calculations show, that the observed fluorescence components with 200 ps in fig. 7, 350 ps in fig. 5 and 600 ps in fig. 11 in the PBP fluorescence (645 – 665 nm) emission are not explained from the coupled system according to fig. 13. The 200 – 600 ps fluorescence components in the PBP fluorescence (645-665 nm) are therefore ascribed to the fluorescence of decoupled PBP pigments which formed new chemical bounds, e.g. PBP aggregates.

The 1.2 – 1.4 ns fluorescence decay component in the Chl d emission (725 nm) results from closed reaction centers and is therefore ignored in fig. 13 [16],[30].

6. Discussion

After excitation with 632 nm laser light the fluorescence dynamics in A. *marina* is characterized by two emission bands which are located at 650 nm and 725 nm (fig. 3). The 650 nm emission is ascribed to PBP fluorescence and the 725 nm band to Chl d fluorescence [16]. A Gaussian fit of the time integrated fluorescence spectrum (fig. 4) after excitation with 600 nm LED light shows that the 650 nm band is composed by two broad emission bands peaking at 644±3 nm and 665±3 nm which are ascribed to phycocyanin (PC) and allophycocyanin (APC) in agreement with other studies [14],[15]. The Gaussian fit further reveals, that the emission above 700 nm can be subdivided into one main band at 725 nm and two smaller Gaussian bands located at 703 nm and 740 nm. The 740 nm fluorescence originates from the PS I [10], while the 703 nm fluorescence is ascribed to the so-called PCB antenna of A. *marina*. The PCB antennae are membrane-intrinsic pigment protein complexes containing Chl d [31] that is connected to the Chl d in the Core antenna of PS II fluorescing at 725 nm.

The time resolved analysis of the fluorescence dynamics in the 645-665 nm spectral range shows that after excitation with 632 nm laser pulses the fluorescence dynamics can be described by three exponential components with lifetimes of 30 ps, 70 ps and 200 ps (fig. 7). The 200 ps component in the PBP fluorescence (645-665 nm) was varying between 200 ps and 600 ps in different experiments (fig. 5, 7, 11). The experimental condition, maybe the preparation and treatment of the sample seem to influence this 200 ps – 600 ps time constant.

The DAS of the 30 ps component has positive amplitude in the PC emission band (645 nm) and negative amplitude in the APC range (655-665 nm) (fig. 7) indicating that the equilibration of the excitation energy between the PC and APC molecules occurs with a time constant of about 30 ps. The positive amplitude of the 70 ps component is found in the whole PBP (645-665 nm) and PCB (703 nm) emission and a negative amplitude in the Chl d fluorescence (725 nm). Therefore the excitation energy is transferred from the PBP antenna to Chl d in the reaction center (RC) of PS II in A. *marina* with a time constant of about 60-70 ps in agreement to earlier studies [16]. This is about 3 times faster than the energy transfer from the PBP to the RC in typical Chl a containing cyanobacteria (e.g. *Synechococcus* 6301) containing larger PBP antennae which are called phycobilisomes (PBS) [32]. In the present study the assumption of such a fast energy transfer from the PBPs to Chl d in the RC of PS II within 70 ps in A.*marina* is supported by the finding that cold stress results in a deceleration of the PBP fluorescence decay (645-665nm) in *A.marina*, which is proportional to the simultaneously occurring decrease of the initial amplitude of the Chl *d* fluorescence (725 nm) (fig. 12).

The results of the present study show, that the excitation energy transfer from the PBP-antenna to PS II is affected by cold stress.

When cells of *A.marina* were exposed to cold stress (i.e. incubation at 273 K) a decrease of the initial amplitude of the 100 ps decay component and a complimentary increase of the contributions of a 600 ps and a 1.4 ns component was observed (fig. 11). The time constant of the fastest component increased during cold stress from 70 ps to 140 ps (data not shown) and is therefore denoted to 100 ps. The amplitude of the 100 ps component which reflects the number of PBP-complexes that can transfer their excitation energy to PS II decreased from 98 % at 298 K to 25 % after 240 min. of incubation at 273 K. This suggests that about 75 % of the PBP antenna systems were decoupled from PS II. This was supported by the increase of the 1.4 ns component during the cold stress. Such long lived fluorescence decay has to be explained by isolated PBP antenna systems [16], [30], [33].

If the decoupling of the PBP-antenna from PS II was caused by a spatial displacement, 75 % of the PBP rods which exhibited 600 ps or 1.4 ns fluorescence decay would have moved during cold stress for a distance large enough to prevent energy transfer. In contrast, the change of the fast time constant from 70 ps to 140 ps would be explained by a very small distance change (+ 13 % of the initial distance) of the remaining 25 % of the PBP rods according to Foerster Resonance Energy Transfer.

If cells of *A. marina* were rewarmed to 298 K after cold stress an increase of the amplitude of the 100 ps decay component was observed again (Fig 11). This finding and the accompanying recovery of the Chl *d* fluorescence intensity upon rewarming suggest that the cold stress induced decoupling of the PBPs is at least partially reversible.

The appearance of a strong varying 200 ps – 600 ps component in fig. 5,7,11 can not be easily explained by a perfectly coupled antenna system according to the model presented in section 5 and is also not typical for decoupled PBP rods [30],[33]. Therefore it is suggested that the observed 200 ps – 600 ps time constant was the decay time of PBP antennae which were conformationally distorted, formed new chemical bonds and/or were in an aggregated state [33].

Such distorted antennae seem to appear during the decoupling. The conformational change could depend on the experimental conditions which would explain the strong variation of the time constant of this decay component from 200 ps to 600 ps.

This assumption would explain the finding that these distorted PBP antennae did not rebind to the Chl *d* containing antenna after cold induced decoupling of the PBP antennae and subsequent rewarming. They were not able to rebind because they formed new chemical bonds. A quantity of PBP rods equal to the relative amplitude of this 200 ps – 600 ps decay component was distorted. Therefore the 200 ps – 600 ps component did not vanish when the stress condition was finished (fig. 11).

The 200 ps – 600 ps decay component could be used for measurements of the fraction of distorted PBP antenna complexes.

A very detailed study by Schreiber [34] has analyzed cold stress induced PBS decoupling which was stimulated by additional illumination. Schreiber observed nearly a full reversibility of the PBS detachment and suggested an accumulation of negative charge on the outer membrane surface responsible for the decoupling. Glazer et al. suggested possible disruption of the coupling of phycobilisomes of Synechococcus to PSI after treatment with N-ethylmaleimide [35] without clarifying the mechanism. Mullineaux suggested a slow membrane protein mechanism leading to a mobility of phycobilisomes (PBS) [27]. Temperature induced decoupling effects of PBS were also observed on Spirulena platensis by Li et al [28].

The effect observed in the study of Li et al. was of smaller amplitude than our results on *A.marina*. Maybe the decoupling of the rod shaped structures in *A.marina* is more effective due to their smaller dimensions in comparison to PBS in other cyanobacteria. In that case not only the energy transfer from the PBP to the RC occurring in 70 ps but also it's regulation by PBP decoupling is more effective in *A.marina* than in other cyanobacteria containing phycobilisomes.

The exact mechanism of the PBP mobility still has to be clarified but seems to occur at different stress conditions. Such decoupling was observed in different cyanobacteria not only during cold stress but also during heat stress or nitrogen starvation (data not shown). It was observed that the decoupling during heat stress was more efficient if the cells were additionally illuminated. This finding was also observed under cold stress by Schreiber et al. [34].

Therefore it is suggested that some stress factors resulting in a lower quenching of excited states in the RC of PS II activates a mechanism for PBP decoupling which reduces the energy transfer to the RC in PS II to avoid photo damage by singlet oxygen generation. For further studies of the PBP movement very short data acquisition times are necessary which are achieved for example using new multi anode systems which are shown in fig. 2b) and which are able to collect up to 4 Million counts/ sec. [22].

Using count rates of several million counts/sec. the fluorescence dynamics can be acquired after < 1Sek. of measurement time. Thus the fluorescence decay could be investigated during even faster metabolic changes.

With the current technique as described here for time resolved fluorescence measurements it is possible to monitor the metabolic change of living cells with ms resolution.

7. References

1. H. Miyashita, H. Ikemoto, N. Kurano, K. Adachi, M. Chihara and S. Miyachi, "Chlorophyll d as a major pigment" *Nature* (London), 383,402 (1996)
2. F.-J. Schmitt, H.-J. Eckert, H.J. Eichler, "Analytical Methods based on fluorescence spectrometry", *Bericht des optischen Instituts,* Berlin, (2005), http://www.physik.tu-berlin.de/institute/OI/schmitt.zip, file: COST_proceedings.pdf
3. Govindjee, "Sixty-three years since Kautsky: Chlorophyll a fluorescence" , *AustJ Plant Physiol.* 22: 131 –160 (1995)
4. K. Sauer, M. Debreczeny *"Fluorescence"* in: J. Auresz and A. Hoff (eds), *Biophysical Techniques in Photosynthesis,* Kluwer Academic Publishers, London, 1996
5. M. Richter, K. Jun Ahn, A. Knorr, A. Schliwa, D. Bimberg, M. El-Amine Madjet, T. Renger "Theory of excitation transfer in coupled nanostructures - from quantum dots to light harvesting complexes" *physica status solidi (b),* 243, 2302-2310 (2006)

6. H. Miyashita, K. Adachi, N. Kurano, H. Ikemoto, M. Chihara and S. Miyachi "Pigment composition of a novel oxygenic photosynthetic prokaryote containing chlorophyll *d* as the major chlorophyll, *Plant Cell Physiol.*, 38, 274-281 (1997)
7. M. Kühl, M. Chen, P.J. Ralph, U. Schreiber and A.W.D. Larkum, "Niche and photosynthesis of Chlorophyll d-containing cyanobacteria", *Nature* (London) 433, 820 (2005)
8. M. Mimuro, K. Hirayama,K. Uezono,H. Miyashita and S. Miyachi, "Uphill energy transfer in a chlorophyll *d*-dominating oxygenic photosynthetic prokaryote, *Acaryochloris marina*" *Biochim. Biophys. Acta* 1456, 27-34, (2000)
9. M. Mimuro, S. Akimoto, T. Gotoh, M. Yokono, M. Akiyama, T. Tsuchiya, H. Miyashita, M. Kobayashi, I. Yamazaki, „Identification of the primary electron donor in PS II of the Chl *d*-dominated cyanobacterium *Acaryochloris marina*", *FEBS letters* 556, 95–98 (2004)
10. M. Mimuro, S. Akimoto, I. Yamazaki, H. Miyashita, S. Miyachi, "Fluorescence properties of chlorophyll d-dominating prokaryotic alga, Acaryochloris marina: studies using time-resolved fluorescence spectroscopy on intact cells", *Biochimica et Biophysica Acta*, 1412 S. 37-46 (1999)
11. E. Schlodder, M. Cetin, H.-J. Eckert, F.-J. Schmitt, A. Telfer "Both chlorophylls *a* and *d* are essential for the photochemistry in Photosystem II of the cyanobacterium, *Acaryochloris marina*", Biochemica and Biophysica Acta, doi:10.1016/j.bbabio.2007.02.018 (2007)
12. D. Shevela, B. Nöring, H.-J. Eckert, J. Messinger, G. Renger, "Characterization of the water oxidizing complex of Photosystem II of the Chl *d*-containing cyanobacterium *Acaryochloris marina via* its reactivity towards endogenous electron donors and acceptors", *Physical Chemistry Chemical Physics*, 8, 3460 (2006)
13. Q. Hu, H. Miyashita, I. Iwasaki, N. Kurano, S. Miyachi, M. Iwaki, S. Itoh, "A Photosystem I reaction center driven by chlorophyll d in oxygenic photosynthesis", *Proc. Natl. Acad. Sci. Plant Biology*, Vol 95 , 13319-13323 (1998)
14. J. Marquardt, H. Senger, H. Miyashita, S. Miyachi, E. Mörschel, „Isolation and Characterization of phycobiliprotein aggregates from Acaryochloris marina, a prochloron like prokaryote containing mainly chlorophyll d". *FEBS Lett* 410, 428-432 (1997)
15. Q. Hu, J. Marquardt, I. Iwasaki, H. Miyashita, N. Kurano, E. Mörschel, S. Miyachi „Molecular structure, localization and function of phycobiliproteins in the chlorophyll a/d containing oxygenic photosynthetic prokaryote Acaryochloris marina", *Biochimica et Biophysica Acta* 1412, 250–261 (1999)
16. Z. Petrášek, F.-J. Schmitt, Ch. Theiss, J. Huyer, M. Chen. A. Larkum, H. J. Eichler, K. Kemnitz and H.-J. Eckert "Excitation energy transfer from Phycobiliprotein to Chlorophyll d in intact cells of *Acaryochloris marina* studied by time- and wavelength resolved fluorescence spectroscopy", *Photochem. Photobiol. Sci.*, 4, 1016-1022 (2005)
17. M. Mimuro "Photon capture, exciton migration and fluorescence emission in cyanobacteria and red algae" in G.C. Papageorgiou, Govindjee (eds), *Chlorophyll a Fluorescence, A signature of photosynthesis,* Springer, Dordrecht, The Netherlands 2004
18. S. Itoh, K. Sugiura "Fluorescence of Photosystem I" in G.C. Papageorgiou, Govindjee (eds) , *Chlorophyll a Fluorescence, A signature of photosynthesis,* Springer, Dordrecht, The Netherlands 2004
19. M. Chen, R.G. Quinnell and A.W.D. Larkum "The major light-harvesting pigment protein of *A.marina*" ,*FEBS Lett.,* 514, 231-250 (2002)
20. A. Bergmann, H.J. Eichler, H.-J. Eckert and G. Renger "Picosecond laser-fluorometer with simultaneous time and wavelength resolution for monitoring decay spectra of

photoinhibited Photosystem II particles at 277 K and 10 K", *Photosynthesis research* 58, 303-310 (1998)

21. A. Bergmann, *Picosekunden-Fluoreszenzspektroskopie mithilfe doppeltkorrelierter Einzelphotonendetektion zur Untersuchung der Primärprozesse im Photosystem II*, Mensch und Buch Verlag, Berlin, 1999

22. Becker & Hickl GmbH, *PML-16-C 16 Channel Detector Head for Time-Correlated Single Photon Counting User Handbook"*, http://www.becker-hickl.de/pdf/pml16c21_.pdf, Berlin, 2006

23. D. Kim, P.T.C. So, "Photon counting by large-area detection with multianode photomultiplier tube in high-throughput multiphoton microscopy", *SPIE Proc. 2006, 6372-12*

24. W. H. Press, A.T. Saul, W. T. Vetterling and B. P. Flannery, "Numerical Recipes in C", Cambridge University Press 1993, www.nr.com

25. Gilmore, "Excess light stress: Probing excitation dissipation mechanisms through global analysis of time- and wavelength-resolved chlorophyll a fluorescence", in: G.C. Papageorgiou and Govindjee (eds) *Chlorophyll a Fluorescence: A Signature of Photosynthesis*, 2004, pp. 555-581. Springer: Dordrecht, The Netherlands.

26. Boguslaw Stec, Robert F. Troxler, Martha M. Teeter, „Crystal- Structure of C-Phycocyanin from Cyanidium caldarium provides a new perspective on phycobilisome Assembly", *Biophysical Journal Volume 76* (June 1999), 2912-2921

27. C. W. Mullineaux, M. J. Tobin, G. R. Jones, "Mobility of photosynthetic complexes in thylakoid membranes", *Nature*, Vol. 390 , 421-424 (1997)

28. Y. Li, J. Zhang, J. Xie, J. Zhao,L. Jiang, "Temperature –induced decoupling of phycobilisomes from reaction centers", *Biochimica et Biophysica Acta* 1504 229-234 (2001)

29. K. Baier, S. Nicklisch, C. Grundner, J. Reinecke, W. Lockau, "Expression of two *nblA*-homologous genes is required for phycobilisome degradation in nitrogen-starved *Synechocystis* sp. PCC6803, *FEMS Microbiology Letters*, 195, 35-39 (2001)

30. F.-J. Schmitt, *"Untersuchung der Fluoreszenzdynamik im Antenensystem des Chlorophyll d haltigen Cyanobakteriums Acarxochloris marina"*, Diplomarbeit im Fachbereich Physik der TU Berlin, Berlin, 2005, http://www.physik.tu-berlin.de/institute/OI/schmitt.zip, file: Diplomarbeit_Franz-Josef Schmitt

31. M. Chen, T. S. Bibby, J. Nield, A.W.D. Larkum, J. Barber, „Structure of a large Photosystem II supercomplex from acaryochloris marina", *FEBS Letters* 579, 1306-1310 (2005)

32. C.W. Mullineaux, A.R. Holzwarth, "Kinetics of excitation energy transfer in the cyanobacterial phycobilisome-Photosystem II complex." *Biochim. Biophys. Acta,* 1098, 68-78 (1991)

33. A.R. Holzwarth, "Structure-function relationships and energy transfer in phycobiliprotein antennae", *Physiologia plantarum* 83: 518-528 (1991)

34. U. Schreiber "Reversible uncoupling of energy transfer between phycobilins and chlorophyll in Anacystis nidulans", *Biochimica and Biophysica Acta,* 591, 361-371 (1980)

35. A. Glazer, Y. Gindt, C. Chan, K. Sauer "Selective disruption of energy flow from phycobilisomes to Photosystem I", *photosynthesis research* 40, 167-173 (1994)

8. About the author

Dipl. Phys. Franz-Josef Schmitt is a scientific coworker in the group of Prof. Eichler in the institute of optics and atom physics of the Technical University Berlin.
He is employed in the SfB 429 "Molecular Physiology, Energetics, and Regulation of Primary Metabolism in Plants" and writing his PhD thesis about primary energy and electron transfer processes in photosynthesis.
The main research fields of Dipl. Phys. Franz-Josef Schmitt are time resolved fluorescence spectroscopic studies of biological samples, fluorescence quantum yield measurements of photosynthetic organisms and combined optoelectronic methods for the visualization of nanoscaled structures (gold nanoparticles in contact with human cells).
In the SS 2007 Dipl. Phys. Franz-Josef Schmitt also supervised a project experiment of the advanced practicum for physics students and lectured the basics of thermodynamics in the Studienkolleg of the Technical University of Berlin.